SPOTLIGHT ON NATURE

FALCON

PAMELA DELL

CREATIVE EDUCATION · CREATIVE PAPERBACKS

Published by Creative Education and Creative Paperbacks
P.O. Box 227, Mankato, Minnesota 56002
Creative Education and Creative Paperbacks are imprints of The Creative Company
www.thecreativecompany.us

Design and production by Blue Design, Inc.
Art direction by Tom Morgan
Edited by Grace Beltowski

Photographs by Alamy Stock Photo/JAH, 21; Dreamstime/Deepgreen, 22, Rsooll, 2–3, 29, Sally Wallis, 15, Steven Oehlenschlager, 16, Wildlife World, 11; flickr/Biodiversity Heritage Library, 8, 10, 14, 16, 20, 22, 26; Getty Images/Javier Fernández Sánchez, cover, 1, W. Perry Conway, 18; Pexels/David Floyd, 23; Shutterstock/Inna Astakhova, 17, kospet, 12, Porojnicu Stelian, 21; Unsplash/Geoffrey Moore, 10, Sangeeth N, 24; Wikimedia Commons/Bernard DUPONT, 29, Dominic Sherony, 29, John James Audubon, 29, jojo nicdao, 28, Marc Dalmulder, 9, Metropolitan Transportation Authority of the State of New York, 4–5, National Park Service, 27, New York City Department of Environmental Protection / Christopher Nadareski, 6

Library of Congress Cataloging-in-Publication Data
Names: Dell, Pamela, author.
Title: Falcon / Pamela Dell.
Description: Mankato, Minnesota : Creative Education / Creative Paperbacks, [2026] | Series: Spotlight on nature | Includes bibliographical references and index. | Audience: Ages 10-13 | Audience: Grades 4-6 | Summary: "Swoop into the falcon's world with this nature title for middle-grade wildlife lovers. Informational text pairs with a narrative about a single falcon family to spotlight the bird of prey's life cycle, supported by infographics and a timeline of developmental milestones"—Provided by publisher.
Identifiers: LCCN 2024048393 (print) | LCCN 2024048394 (ebook) | ISBN 9798889896203 (library binding) | ISBN 9781682777862 (paperback) | ISBN 9798889897002 (ebook)
Subjects: LCSH: Peregrine falcon—Juvenile literature.
Classification: LCC QL696.F34 D45 2026 (print) | LCC QL696.F34 (ebook) | DDC 598.9/6—dc23/eng/20241223
LC record available at https://lccn.loc.gov/2024048393
LC ebook record available at https://lccn.loc.gov/2024048394

Printed in India

CONTENTS

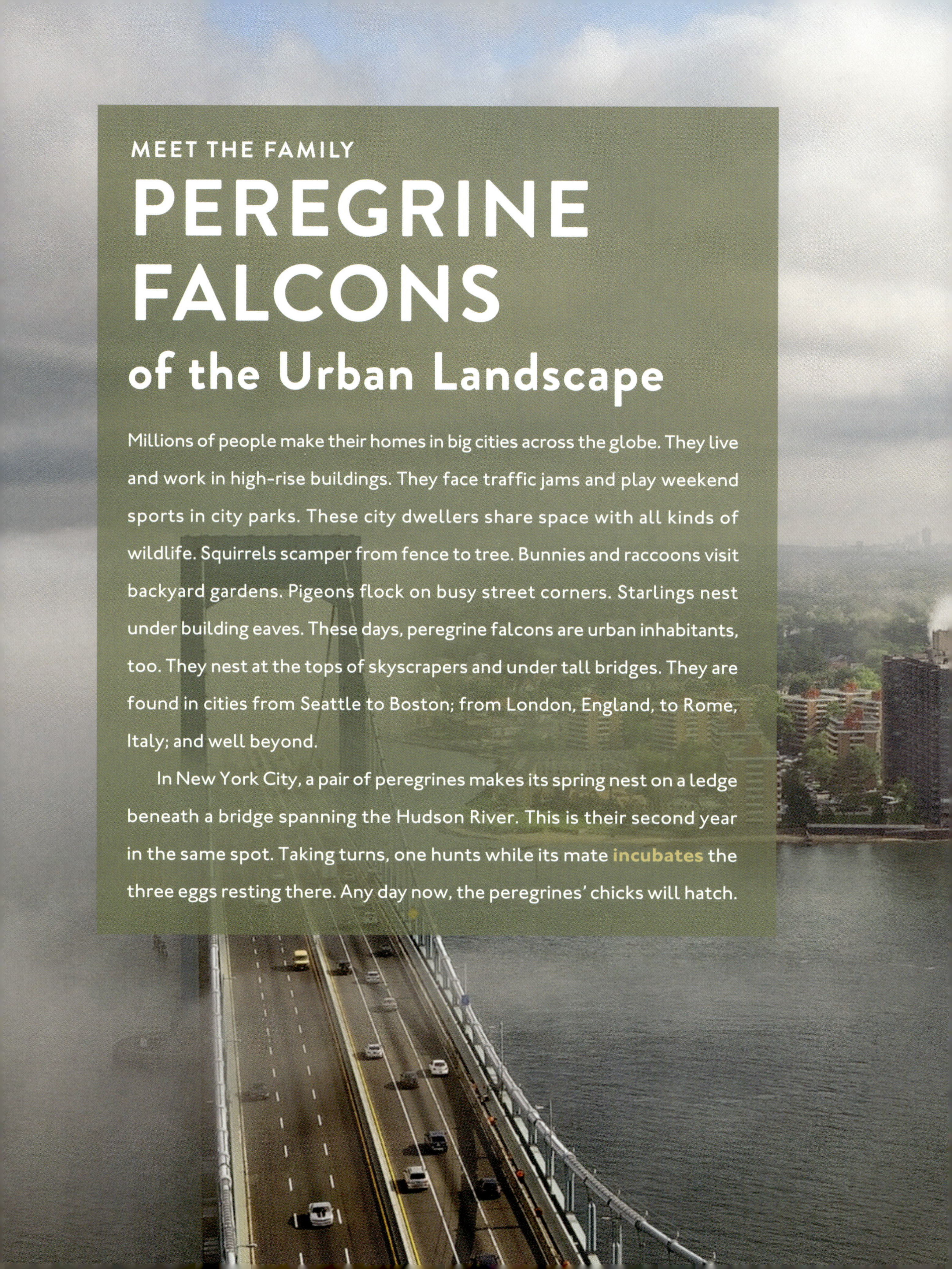

MEET THE FAMILY

PEREGRINE FALCONS

of the Urban Landscape

Millions of people make their homes in big cities across the globe. They live and work in high-rise buildings. They face traffic jams and play weekend sports in city parks. These city dwellers share space with all kinds of wildlife. Squirrels scamper from fence to tree. Bunnies and raccoons visit backyard gardens. Pigeons flock on busy street corners. Starlings nest under building eaves. These days, peregrine falcons are urban inhabitants, too. They nest at the tops of skyscrapers and under tall bridges. They are found in cities from Seattle to Boston; from London, England, to Rome, Italy; and well beyond.

In New York City, a pair of peregrines makes its spring nest on a ledge beneath a bridge spanning the Hudson River. This is their second year in the same spot. Taking turns, one hunts while its mate **incubates** the three eggs resting there. Any day now, the peregrines' chicks will hatch.

CLOSE-UP

Egg tooth

A falcon chick's **egg tooth** is not a real tooth. It is a growth on the baby bird's beak used to break out of its egg. Several weeks after the chick pips through the egg, the egg tooth falls off.

CHAPTER ONE

LIFE BEGINS

The falcon family includes more than 60 **species**. These species are divided into a few groups, not all called falcons. Some are known as caracaras and kestrels. Also part of the falcon family are falconets, hobbies, and the gyrfalcon (*JUR-fal-kn*).

Falcons live on every continent except Antarctica. Some species live in dense forests. Others make their homes in deserts, wetlands, or high-mountain regions. All falcons are **raptors**. They eat meat, including other birds, rodents, fish, and insects.

Falcons are also prey themselves. Owls and other birds of prey hunt them. Big cats and foxes are falcon enemies. But strong adults usually aren't the target. **Predators** attack small, weak, or young falcons first.

In almost all falcon species, the coloring of both males and females is usually brown or dark gray with lighter underparts. But, like most raptors, female falcons are slightly larger. Falcons range in size depending on their

PEREGRINE FALCON MILESTONES

2 DAYS

- Completes pipping, free of shell
- Weight: 1.5 ounces (42.5 grams)

6 DAYS

- Parents take turns bringing food
- Weight: 3 ounces (85 g)

species. They may be only a few inches (centimeters) long or as large as 2 feet (0.6 meter). A falcon's weight can range from a few ounces (grams) to 4 pounds (1.8 kilograms).

Once they are independent, falcons stake out their own territories. Some species work hard to defend these hunting spaces from others of their kind. They go easier on birds of other species. Most falcons mate for life with the same partner. Then, defending the territory also means defending the nesting area. If one of the pair dies, the other finds a new mate.

Falcons build nests that are unlike those of most birds. When a mother falcon prepares her nest, she does it simply. She doesn't carry branches or twigs back to her nesting spot. She lies down and

CLOSE-UP

Coloration

Most falcons have brown or dark gray **plumage** on their backs, with mostly white or lighter-colored underparts. Juvenile falcons are similar in color to the adults, with a lot of streaking in their underparts. Peregrine falcons' feathers are more bluish-gray than brown.

FEATURED FAMILY

Welcome to the World

Beneath the bridge is a metal ledge where the peregrines' nest lies. In the nest, just sitting on the ledge's rubble, are three reddish-brown speckled eggs. A female falcon watches over the eggs. She rolls and pushes them so the chicks inside will develop properly. One spring day, a pip, or tiny hole, appears in one egg, then in the others. Each chick uses its egg tooth to break the shell. As they struggle, their mother stays close. She chirps to encourage them. Within two days the chicks escape. Now, a new world is waiting for them.

pushes backward with one foot. This creates a shallow indent, or "scrape," in the pebbles or gravel. This is her nest. Many falcons, like peregrines, nest on high rock ledges or other places far above the ground. Some species nest in tree hollows. Sometimes falcons simply take over the abandoned nests of other birds.

Falcon chicks are called eyases. They come into the world helpless. It will be weeks before the eyases **fledge**. Until then, their parents guard them. They take turns bringing food back to the nest. They work full time to keep their **hatchlings** alive.

2 WEEKS

- Egg tooth has fallen off
- Parents continue to bring food
- Left alone for extended periods

CLOSE-UP

Brood patch

To help keep their eggs warm, incubating falcons develop a brood patch. This featherless area of a female's belly has thickened skin and greater blood flow. This passes warmth into the eggs. Male peregrines also develop brood patches, but they are less pronounced.

FEATURED FAMILY

First Meal

The three peregrine eyases have just hatched. Covered in down feathers, the hatchlings look like little balls of white fluff. Their feet are awkward and large compared to their bodies. And they are hungry. Before they hatched, their parents shared incubation duties. One kept the eggs warm while the other hunted. But now the parents take turns hunting to feed their chicks. They return to the nest with chunks of meat. Using their sharp beaks, they tear off bits of the meat. They carefully feed their open-mouthed, begging little eyases.

Falcon chicks come into the world

HELPLESS.

Their parents take turns bringing food back to the nest.

③ WEEKS

- Able to walk
- Weight: 15 ounces (425 g)

⑤ WEEKS

- Juvenile feathers begin to replace downy fluff

CLOSE-UP

Keen vision

Scientists believe falcon eyesight is about eight times sharper than humans'. They can spot prey from 2 miles (3.2 kilometers) away. They have two yellow eyelids, plus a see-through third one. This eyelid protects and moistens their eyes, especially during flight.

CHAPTER TWO

EARLY ADVENTURES

Falcon pairs mate and raise a family once a year. This usually takes place between late winter and early summer. During this time, there is more prey to feed the adults and their chicks. The female normally lays two to four eggs, sometimes more. In about four to five weeks the eggs will hatch. All the eyases emerge around the same time, so they will grow at the same rate. Because of this, there won't be smaller, weaker chicks that could be killed by their siblings.

For the first two or three weeks, one parent will hunt while the other sits in the nest keeping the **nestlings** warm and protected. In their first days, the chicks are weak. They cannot even stand up. They scrabble around the nest, crying for food. They can't regulate their own body temperatures either. They need the warmth and shelter of a parent's body to keep them alive. This is especially true at night.

6 WEEKS

- Fully feathered
- First attempts to fly

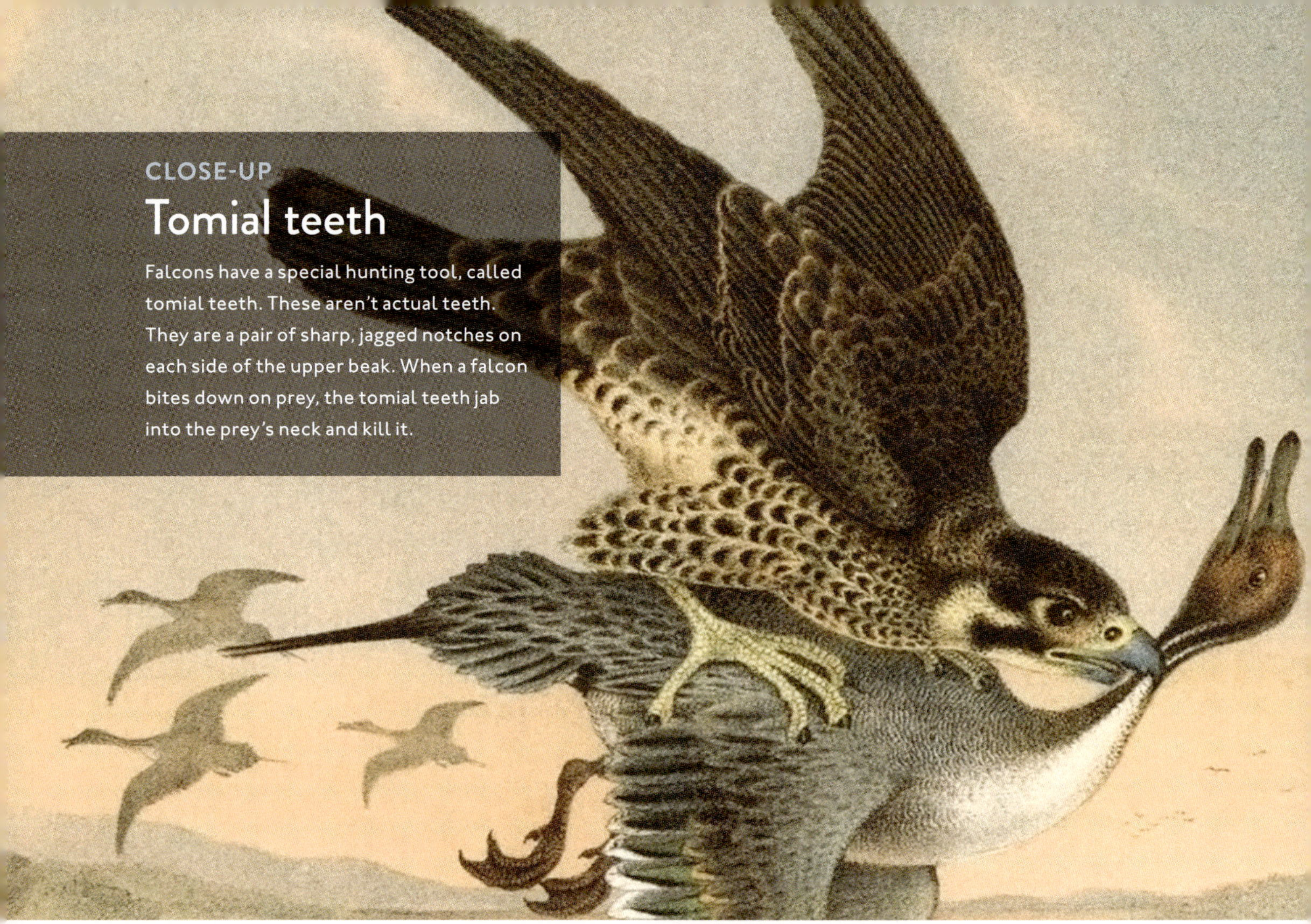

CLOSE-UP

Tomial teeth

Falcons have a special hunting tool, called tomial teeth. These aren't actual teeth. They are a pair of sharp, jagged notches on each side of the upper beak. When a falcon bites down on prey, the tomial teeth jab into the prey's neck and kill it.

FEATURED FAMILY

Getting the Hang of It

For several days, the three little falcon chicks have been practicing hopping around the nest and flapping their wings. Now they are about six weeks old. They have all fledged, or begun to fly. But just barely. They don't go far from the nest. They perch on a nearby ledge, squawking loudly. They're begging their parents for food. The young falcons drag their wings along the ledge. They seem to be tilting and wobbling and almost toppling over. But they're not in trouble. This is just a normal part of learning to fly well.

The chicks grow rapidly. Gradually, the parents leave them on their own for longer periods of time. In three to five weeks, harder feathers begin to replace the downy fluff. After meals, the young falcons start preening—using their beaks to clean and straighten the new feathers. They explore the area around the nest. They also spend a lot of time flapping their wings and taking short hops. These are their early attempts at flight. And it is good exercise for strengthening the falcons' wings. They will soon be flying.

10 WEEKS

- Able to hunt on its own

15 WEEKS

- Able to fully care for itself

CLOSE-UP

Stooping

Peregrine falcons are speed demons. When chasing prey, they can fly about 60 miles (96.5 km) per hour. Stooping is the act of diving straight down to attack prey. During a stoop, falcons may reach speeds of up to 240 miles (386 km) per hour.

FEATURED FAMILY

Give It a Try

One of the peregrine falcons has recently fledged. His wings are growing strong. But he is not yet totally independent. His hunting skills are just developing. He sits on a backyard pine bough, alert and watchful. He is small. Many predators would still be eager to make a meal of him. But he is a predator, too. He spots a small finch on a nearby fence and goes for it. He's fast, but the bird escapes. The young peregrine accidentally hits the fence. He's unharmed, but he still has a lot to learn about hunting.

AT THREE TO FIVE WEEKS OLD,

harder juvenile feathers begin to grow in, replacing the young falcons' downy fluff.

13 MONTHS

- Sheds feathers for first time
- Considered an adult
- Weight: 1–3.5 pounds (0.5–1.6 kg)
- Wingspan: 39.4–43.3 inches (100–110 cm)

CLOSE-UP

Talons

Falcons have four long, razor-sharp claws, called talons. Three face forward and one faces backward. Falcons usually catch other birds that are in mid-flight. The talons reach through the prey's feathers and clutch tightly. Curled fist-like, the talons can also knock prey out of the air.

CHAPTER THREE

LIFE LESSONS

The lifespan of many wild falcon species is difficult to know. Some live in thick forests or in isolated areas. It is hard to track individual birds in these places. But falcons are some of the longest lived of all birds. Smaller falcons have shorter lifespans than larger ones, such as peregrines. Peregrines can live as long as 15 years. The smallest of all falcons, the pygmy, is known to live about eight years. Captive birds generally have longer lifespans than those in the wild.

The first year of life is the most dangerous for falcons. If the eggs don't break and the eyases hatch, the struggle is on. About 6 out of 10 die before they are a year old. Cold weather or other natural forces could do the chicks in. They may become the victims of horned owls or other predators. But if they make it through that first year, their luck increases. They become much more likely to survive and raise their own families.

Falcons fledge at different rates. American kestrels leave their nesting holes and begin practice flights at about three weeks. Gyrfalcons first try

2 YEARS

- Finds a mate
- Ready to nest

out short flights at five weeks. Pergrines fledge at around six weeks but stay near their parents. Merlins are totally independent by three months. No matter when they fledge, these young birds need a lot of practice flying.

Once they are successful fliers, young falcons follow their parents through the sky. They observe the adult birds in action, hunting airborne prey. This shows them how to carry out their own hunts. Unlike many raptors, falcons don't use their talons, or claws, for killing. Talons hold prey. Falcons kill with their beak and tomial teeth.

The ability to fly is **inborn**. But parents may provide their own lessons. They speed through the air holding captured prey. The young bird follows until it catches up. If successful, it will grab the prey from its parent's talons right in flight. Adults may also catch prey without killing it. Then they release the catch so their young can kill it themselves. These valuable exchanges ensure that young falcons are skilled to meet the future.

FEATURED FAMILY

This Is How It's Done

The peregrine falcon is improving his hunting abilities. He flies well by now. But today he waits, perched on the ledge of a skyscraper. His mother flies near, holding a large chunk of meat in her beak. The peregrine sails toward her, calling for food. His mother shares. But the youngster is still hungry. He swoops away, keen eyes watchful. Soon, a small songbird catches his attention. He speeds up and gives chase at 60 miles (96.5 km) per hour. The smaller bird has no chance. The peregrine is becoming a super-skilled hunter.

CLOSE-UP

Feathers

Falcon wings are generally narrow and pointed. Some species have slightly more rounded wings. Each wing has 22 main feathers. A falcon's leg feathers, or plumes, are called flags. Molting, the shedding of old feathers for new, can take months to complete.

(3) YEARS

- Fully grown
- Females lay two to four eggs each spring

CLOSE-UP

Falconry

Falconry is a 4,000-year-old sport. Falconers train falcons to hunt for game, such as rabbits. In falconry, falcons also help control mice and other pests. Peregrines and American kestrels are easily trained for this sport. Strict U.S. regulations help protect the birds.

FEATURED FAMILY

Practice Makes Perfect

The growing peregrine soars past the edge of town, moving at 40 miles (64.3 km) per hour. The wind is up, and he is hungry. Far below, a duck rises from the river. The falcon folds his wings back and dives, or stoops, straight downward. He moves at a breakneck speed that tops 200 miles (322 km) per hour. With his feet curled fist-like, he strikes the duck. The duck falls, stunned. The peregrine catches and kills it with his tomial teeth. His first successful stoop!

To help their young learn to hunt,

ADULT FALCONS

release prey to them in midair.

15 YEARS

- End of life

CHAPTER FOUR

HOPE FOR THE FUTURE

Today, most falcon species are not considered endangered. But in the early 1970s, the North American falcon population was in serious decline. The peregrine falcon in particular was barely holding on. Across the continent, their population had mysteriously dropped. The peregrine was already **extinct** east of the Mississippi River. By 1975, only 324 pairs of nesting peregrines could be accounted for. European countries faced similar issues.

The cause, scientists discovered, was DDT. DDT is a pesticide, a product sprayed on crops to kill insects that feed on those crops. Falcons and other birds were eating the animals that ate the poisoned insects. The birds' bodies became full of DDT, too. It wasn't just grown falcons being killed either. Because of pesticides, falcon eggshells were too thin and fragile. Before a baby chick could even develop inside, the egg would collapse. Very few new chicks were being added to the falcon population.

Widespread public awareness campaigns brought the dangers of pesticides to people's attention. In 1972, DDT was almost completely banned in North America and much of Europe. Through these efforts, falcons soon began to make a comeback. By 1999, the number of North American peregrines had increased to 1,650. This success removed them from the endangered species list.

This is good news, but falcons still face many threats. Experts believe raptors as a group are in decline. The biggest problem is loss of habitat. People move into falcon territory and take over. As land is cleared for human use, birds lose nesting sites and prey. Climate change poses another danger. The gyrfalcon, for instance, needs very cold weather to survive. If the arctic climate warms up too much, they'll be gone. Many falcons also die of starvation or from collisions with human-made objects.

Today, important efforts are being made to increase falcon populations worldwide. The Peregrine Fund leads the way in efforts to keep raptors from extinction. Another resource is the National Park Trust. Their website provides pointers for anyone who wants to help keep falcons safe. The careful breeding of captive falcons is also helping. Every bit of this work is important. But the first step in helping these swift and beautiful birds survive is becoming aware of the dangers they face.

FAMILY ALBUM

SNAPSHOTS

The **gyrfalcon**, found in the Arctic, is the largest falcon. Females weigh about 4 pounds (1.8 kg), with males weighing slightly less.

The **American kestrel** is the smallest North American falcon. No bigger than a blue jay, it is sometimes called a sparrow hawk.

Most falcon species hunt in broad daylight. But dawn and dusk are the hunting hours for the **Oriental hobby**.

Red-throated caracaras prey on wasps' and bees' nests for food. They don't get stung because they produce a powerful insect repellant.

The **yellow-headed caracara**'s diet includes ticks. It's an easy meal. The birds pick their prey off the backs of cattle.

Only **crested caracaras** use nest-building materials, such as sticks and vines. The bulky nests are usually built in cabbage palm trees.

The **collared forest-falcon** is not endangered yet. But loggers targeting the bird's favorite nesting tree, the Spanish cedar, could mean trouble.

Cassini peregrines are a rare subspecies noted for their pale coloring. They live only in South America's Patagonia region.

Most falcons hunt alone. But **aplomado falcon** pairs work together to flush out and catch small birds.

Amur falcons migrate a record-breaking 9,000 miles (14,484 km) over open water. They fly from Russia and China to eastern and southern Africa.

The **New Zealand falcon** is sometimes called the quail hawk or bush hawk. New Zealand is the only place it lives.

Two of the known longest-lived falcons were a **peregrine falcon** and a **crested caracara**. Both lived at least 22 years.

WORDS to Know

egg tooth a small, sharp knob on the side of a bird's beak that helps it break out of its egg

extinct no longer in existence

fledge to have the muscles and feathers necessary for flight

hatchling an animal that has just recently broken out of its eggshell

inborn present from birth

nestling a newly hatched bird, too young to leave the nest

plumage the entire feathery covering of a bird

predator an animal that hunts other animals

raptor a bird of prey, with a sharp, curved beak and sharp claws, called talons

species a group of living things that have shared characteristics and are able to reproduce with one another

LEARN MORE

Books

Lukidis, Lydia. *Falcons*. Mankato, Minn.: Black Rabbit Books, 2023.

Pallotta, Jerry. *Falcon vs. Hawk*. Minneapolis: Spotlight, 2020.

Riggs, Kate. *Falcons*. Mankato, Minn.: Creative Education and Creative Paperbacks, 2023.

Websites

Jewsbury, Keaton. "How You Can Help Save Birds of Prey." National Park Trust. https://parktrust.org/blog/how-you-can-help-save-birds-of-prey/

"Peregrine Falcon." BioKIDS: Kids' Inquiry of Diverse Species. https://biokids.umich.edu/critters/Falco_peregrinus/

"Peregrine Falcon." National Geographic Kids. https://kids.nationalgeographic.com/animals/birds/facts/peregrine-falcon

Documentaries

Baxter, Simon, dir. *World's Fastest Animal*. Alexandria, Va.: PBS, 2018.

Chabloz, Vincent, dir. *Peregrine Falcon: Lord of the Skies*. Armonk, N.Y.: MagellanTV, 2020.

Izenberg, Joshua, and Brett Marty, dirs. *Game Hawker: A Wild Journey to Falconry*. San Francisco: Speculative Films, 2022.

Note: Every effort has been made to ensure that any websites listed above were active at the time of publication. However, because of the nature of the Internet, it is impossible to guarantee that these sites will remain active indefinitely or that their contents will not be altered.

Visit

AUDUBON CENTER FOR BIRDS OF PREY

See live falcons and other birds of prey on a daytime visit, or sign up for raptor summer camp.

1101 Audubon Way
Maitland, FL 32751

NATIONAL AVIARY

Check out the Peregrine FalconCam or hold Horus the Lanner falcon on your gloved hand.

700 Arch Street
Pittsburgh, PA 15212

DEVIL'S COURTHOUSE

Hike up to this spectacular overlook for a chance to see wild peregrine falcons, known to nest on the surrounding cliffs.

Blue Ridge Parkway Milepost 422
Asheville, NC 28803

WORLD CENTER FOR BIRDS OF PREY

Take tours and make sure to catch the Fall Flight bird show to see several swift falcons in action.

5668 West Flying Hawk Lane
Boise, ID 83709

INDEX